CALGARY PUBLIC LIBRARY

SEP -- 2011

THE BEAVER MANIFESTO

The Beaver Manifesto

Glynnis Hood

RMB

Victoria Vancouver Calgary

Copyright © 2011 Glynnis A. Hood

All rights reserved. No part of this publication may be reproduced, stored in a retrieval system, or transmitted in any form or by any means—electronic, mechanical, audio recording, or otherwise—without the written permission of the publisher or a photocopying licence from Access Copyright, Toronto, Canada.

Rocky Mountain Books
www.rmbooks.com

Library and Archives Canada Cataloguing in Publication

Hood, Glynnis A.
    The beaver manifesto / Glynnis A. Hood.

Includes bibliographical references.
Also issued in electronic format.
ISBN 978-1-926855-58-5

    1. Beavers. I. Title.
QL737.R632H66 2011   599.37   C2011-903299-6

Printed and bound in Canada

Rocky Mountain Books acknowledges the financial support for its publishing program from the Government of Canada through the Canada Book Fund (CBF) and the province of British Columbia through the British Columbia Arts Council and the Book Publishing Tax Credit.

BRITISH COLUMBIA ARTS COUNCIL
Supported by the Province of British Columbia

Canadian Heritage   Patrimoine canadien

The interior pages of this book have been produced on 100% post-consumer recycled paper, processed chlorine free and printed with vegetable-based dyes.

MIX
Paper from responsible sources
FSC® C016245

Contents

Acknowledgements

The comment "a book doesn't write itself" is true, but to transform that writing into meaningful text sometimes takes a village. I am so grateful to my dear friends Dee Patriquin and Dr. Patricia McCormack and my mother, Kathleen Hood, for checking my logic, grammar and flow of ideas. Dee Patriquin provided excellent suggestions and thoughtful guidance throughout this process. She has offered unwavering support, even when I used the word tenacity in a most tenacious manner. Patricia McCormack's extensive expertise in the study of the North American fur trade provided an invaluable assessment of the accuracy and representation of those and many other aspects of the book. Her suggestions throughout the entire manuscript were most welcome. Kathleen Hood kindly offered comments that aided the flow and understanding of the text, even though it wasn't her preferred genre of the mystery novel. Next time I might write a murder mystery that involves a beaver, a trapper and a rogue moose, just for her. Of course, the inspiration for the research that led to much of the focus and intent of this book was done in collaboration with my mentor and friend Dr. Suzanne Bayley. She encouraged me to pursue ideas and research approaches that opened new gateways to old ideas. I thank her for her trust and foresight. She continues to inspire me. Wes Olson has always provided his insights and opened the door to the world of beaver surveys, aspen forests and the wonderful ecology of the prairies. I also thank my publisher, Don Gorman, who, after hearing a radio interview I had with the CBC one day in June 2009, envisioned a book from my words. His support and belief in this project are truly appreciated. As an editor, Joe Wilderson brought out the best in the book in a masterful way. Thank you. Finally, I am inspired by a furry rodent with an overbite and a history the size of a country. Who knew that such an odd little creature could make such a big difference to so many?

Prologue

While touring around the province of Quebec one autumn in the mid-1980s, I made friends with a young traveller from France whose only wish was to see a Canadian beaver. At one point, our explorations took us to the lovely village of Tadoussac, at the confluence of the St. Lawrence and Saguenay rivers. When we asked the locals if there was a lodge nearby, they told us to walk a short distance to the edge of the village, where we would find a beaver pond. They guaranteed us that we would see a beaver, and they were right – almost.

We waited on the bank of the small pond. And waited. And waited some more. Eventually Anne-Catherine got up, turned her back to the pond and said, "Enough." At that moment, the beaver rose to the surface of the pond, looked me straight in the eye and I swear it winked. In the second it took Anne-Catherine to turn around, the beaver was gone in a trail of bubbles, never to be seen again. I should have known something was up and that beavers and I were going to have a long and interesting acquaintanceship.

My almost 25-year career working in various protected areas, from the marshlands of the Creston Valley in British Columbia, to the West Coast, the Rockies, the Subarctic and finally the Boreal Plains, has inspired adventures with these furry rodents at almost every turn. Even after I left Parks Canada in 2007 to work at the University of Alberta, the adventures continued with these fascinating but controversial creatures.

As much as beavers have played a role in my personal journey while working in wildlife and resource management, their impacts quickly extend beyond the individual. Beavers are a fitting symbol of the opening up of the North American continent to global integration and the eventual development of an environmental ethic. In some regards they are like the nagging memory that you just cannot shake. That bit of personal history that still has lessons to teach. The fact that beavers still exist in North America, and Europe for that matter, is nothing short of a miracle. Their tenacious nature and ability to survive major geologic and climatic shifts is amazing in its own right, but their ability to survive a level of overexploitation that would have crippled many other species is a testament to their ecological adaptability.

It is this very adaptability that has left humans chasing beavers for hundreds of years: for food, for furs, and now in frustration as these busy engineers flood roadways and other modern landscapes. Over the years, I have seen how coming to understand the very traits of beavers that frustrate people the most can lead to unique opportunities to work with landscapes in a whole new way. It is an opportunity to look at the history of this species, starting long before the North American fur trade, beaver hats, and the expansion of settlement and clearing of land that has changed so much of the ecology of this country, and to find clues within that history to aid our progress. The beaver is a species that has survived both ecological and human challenges in spite of and because of its biology.

Once, while conducting research on beaver foraging ecology in Elk Island National Park, I stepped into a metre-deep vertical hole that had been excavated by a beaver from the bottom up. I never saw it coming because the entrance was covered with a swath of grass. As my foot went deeper into the hole, a stick halfway down twisted my ankle one way and then another stick a little farther down twisted it back again. Using my ankle was hopeless, and it was only with the help of my dog Cass that I was able to get back to the road. As I slowly crawled and limped along the edge of the pond, balancing my weight on Cass's back, I had a lot of time to think. Together, with Cass acting as my crutch, we manoeuvred through a landscape of deep channels, fallen logs, flooded shorelines, more vertical holes, slick foraging trails and finally the dam itself. It was a slow-motion exploration of the remarkable engineering skills of this overgrown rodent. Not only can beavers change the condition of an ankle, they can change the state of the environment, and for the love hats we almost lost it all.

Throughout my career I have seen the sometimes expensive damage beavers can leave in their wake. Many have tried to outwit a 30-kilogram genius; some innovative and patient land managers have developed ways to help humans live with beavers and reap the benefits of the creatures' handiwork. The old resource management techniques still predominate, but given the stream of conversations, phone calls, emails and letters I receive regularly, there is desire for a change in approach. It took the dry years, those without beavers and their ability to put water back on the land, for people to truck, train and even fly beavers back into their old haunts. In most places, a world without beavers is a world without water and the life it supports. In a time when water consumption by humans is at an all-time high and droughts are becoming all too common, the beaver offers us an undeniably "green" solution. We really need a water superhero. As a species that has survived global climate changes over millions of years, perhaps the beaver is just the ally we need.

Beavers truly are the shapers of the physical and ecological landscapes of North America. They and these landscapes have evolved together for millennia and much of our natural environment has really only known a world that has beavers in it. That beavers were almost extinct in most parts of North America just 100 years ago seems remarkable now. Without them, one can imagine the ecological desert left behind. Since sitting by that beaver pond in Quebec so many years ago, I have grown to understand how this species – born from both water and land – has shaped us all. Despite the conflict, the frustration and even fascination, the beaver must be regarded as a superior survivor of a hostile world. It is the waterworks specialist that has helped form and influence the hydrology of our continent for millennia; it is the forester that opens a path of light to the forest floor that inspires new life; and it is one of the only species (other than humans) that can modify landscapes on such a broad scale. So many species are drawn to the beaver's benefits and would perish without them. There is a reason why visitors to Canada seek out a glimpse of a real Canadian beaver: it is an iconic symbol of something inside us that speaks through our history and offers us hope for the future. Just as so much of our ecological landscape was born through beavers, our country was too. Even deep in the heart of our cities we are still part of a world they helped to create.

Beavers on Ice

If I could design the perfect animal, it would be the beaver – even its looks are compelling. Born of an era long before the Last Ice Age, it has since survived cycles of global climate change, sabre-toothed cats, the rise and fall of the Roman, Ottoman and British empires, and now an unprecedented consumption of specialty coffees and reality TV. Its tenacity and environmental ingenuity are matched only by those of one other species: *Homo sapiens sapiens*. Beavers are the great comeback story, a species that outlasted the Ice Age, major droughts, the fur trade, urbanization and near extinction. It is one of the few animals that can go head to head with humans and win. And so the battle for world domination begins.

Although I have been studying beaver ecology for over ten years now and had worked with them long before that, my research represents a mere blip in time relative to the millions of years that some form of beaver has been digging, moving and chewing its way across the planet. Ever since the first rodents came on the scene, between 54 million and 38 million years ago, there has been some form of beaver-like animal waddling its way toward the current vision of Canada and its unique history. The fossil record reveals at least thirty different genera in the beaver family, *Castoroidae*. Yet, only two species remain: the European beaver (*Castor fiber*) and the North American beaver (*Castor canadensis*). Both have been around for at least 1.9 million years, but were almost eliminated after only a few hundred years of hunting, trapping and a quest for really tall fur-felt hats.

By the time early hominid ancestors appeared, just short of 5 million years ago, early beavers had already learned to cut wood and build homes. Beaver Pond, a site on Ellesmere Island in Canada's High Arctic (78° north latitude), dates back 4 million to 5 million years and is littered with cut sticks and fossil remains of a now extinct form of beaver that was about a third the size of the ones we see today. To argue that beavers had a head start on humans is an understatement. Primitive humans, such as Neanderthals, did not start hitting their proverbial stride until 250,000 years ago, and modern humans did not evolve until the Great Leap Forward, a mere 40,000 years ago. As biogeographer Jared Diamond describes in his book *The Third Chimpanzee*, this was a time of accelerated evolution of language and art and a gradual dispersal of humans across the continents. In many ways they followed the path of the beaver into more northern climates and newly formed continents.

In North America, not only were there very small forms of beavers that weighed less than 1.5 kilograms, there was at least one giant that lived among them that ranged up to 3 metres long. The best known is the giant beaver, *Castoroides ohioensis*, which left fossil remains as far south as Florida and as far north as Old Crow,

Yukon. At weights of 60 to 100 kilograms, it was the size of a female black bear, and fortunately for humans, was vegetarian. Beside it was the same beaver we see today – the one that, unlike its bigger cousin, avoided the mass extinction of large mammals at the end of the Last Ice Age. When I am invited to speak at universities, public events and private functions, I often bring along two skulls – a replica one of a giant beaver and one from a beaver of today. Needless to say, the contrast between one thing the size of a medium watermelon and another the size of a small grapefruit captures people's attention and elicits the question: whither *Giant* and why *Little Guy*?

How could this little, buck-toothed, flat-tailed, furry rodent survive when so many other species failed? How could it beat a dramatically changing climate when earlier species of humans and other mammals were permanently laid to rest in the peat deposits and sediments of the past? I would like to say it was due to one key trait, but as with its success today, it was several.

As a semi-aquatic animal, the beaver is a product of the best of both worlds. In some ways it seems like its hind end, with its webbed feet and flat tail, was made for water, while its front end, with its grasping "hands" and chisel-like teeth, is a product of the land. A closer look reveals that water is the beaver's true element. Special valves close the entrances to its ears and nose (every swimmer's dream), and a special membrane covers its eyes like swim goggles so it can see underwater. Even specialized eye muscles correct for refraction so that underwater objects do not look distorted. And finally, its fur-lined inner lips close behind its teeth so it can chew underwater in comfort. All of these sensory organs line up perfectly so they all remain above water as the beaver swims, allowing it to keep watch for predators.

Ranging from the Gulf of Mexico (and prior to the fur trade, even farther south into Mexico) to the Arctic Ocean, the beaver is an animal that can adapt to environmental extremes. Within its northern range, where winter can last for eight to ten months of the year, its coat – a combination of thick underfur and longer guard hairs – protects it against even the coldest waters. When winter covers its ponds with ice, whole families of beavers still swim under the frozen surface from the underwater entrance of their lodges and dens to access their food caches and other underwater vegetation. While other animals walk on top of the ice-covered pond, there is likely something swimming beneath. And just as with seasonal change, the beavers have adapted to a range of major climatic shifts over their evolutionary lifetime.

Although difficult to track at the further reaches of the geologic time scale, cooling and warming periods have been common over the eons. During the Era of Modern Beaver (about 1.9 million years ago to the present), these climatic events are a little more obvious. After the warmer temperatures of the Jurassic Period, the earth started its gradual trend toward cooler temperatures and the evolution of sensibly dressed mammals and birds. Although lush forests still dominated what is now the High

Arctic, the Age of Mammals had begun. The tide really turned, however, when we entered a time better known as the Last Ice Age (Pleistocene), and sheets of ice covered much of the northern hemisphere, approximately 2 million years ago. It marked a period when both human and beaver evolution took off – a real "make it or break it" period in biogeography.

Paleontologists, such as Dr. Natalia Rybczynski from the Canadian Museum of Nature, hypothesize that modern beavers evolved from ancestors that dug burrows, swam and cut down trees. These three characteristics are the secret to the beaver's success and have allowed them to rule the world for millennia – at least until humans made the Great Leap Forward and found language, a sense of adventure and the fashion industry. Some of the predecessors of the modern beaver were better at certain activities than others, but in the end the beaver evolved to become what we see today: a semi-aquatic mammal with a bent toward environmental engineering. This is the very attribute that keeps my phone ringing with various people seeking advice and solace, from frustrated landowners to those who would do just about anything to restore beavers to their rightful place in the ecosystem.

It was the difficulties caused by the cooling temperatures leading into the Last Ice Age that allowed beavers to become magnificent. To endure what are now typical northern winters, caching food was essential. Life under the ice meant life without takeout, fast food or delivery. Planning in advance by cutting stems and storing them at one's front door was not only a necessity, it was an art. Dr. Rybczynski suggests that caching by beavers coevolved with harsh winters and frozen water bodies. To increase the odds of surviving months of ice and snow, beavers also learned to build dams and dig channels that would deepen and expand their ponds so that the food caches could still be accessible underwater from the entrances of their lodges or dens. In the snow-free months, the great expanse of their pond provided not only easy transport of stems back to their lodge without wasting precious energy, but also safe avenues for juvenile beavers to disperse. Their semi-aquatic lifestyle also allowed them to live in a castle (lodge) surrounded by a moat (pond), which protected them from many predators. For good measure, they would also leave along the shoreline the spear-pointed stumps of cut shrubs and trees that could easily impale a clumsy intruder. It would be a long time before the wealthiest of humans would mimic how beavers lived, so they could fend off invading armies and unwelcome neighbours. It would be even longer before they could alter ecosystems to the same degree as this remarkable hydrological engineer.

By the time the Pleistocene ended, almost 10,000 years ago, many of the other mammals had become extinct. Giant ground sloths, mastodons, woolly mammoths, the sabre-toothed cat and even an earlier version of the horse were but a memory. It was time to enter a more modern era – a time of multiple droughts, the Little Ice Age, the nearly fatal fur trade and the incredible ingenuity that allowed beavers to survive it all.

Marketing Kanata

How do you make a country? Hop on a boat, sail across an ocean and chase a furry rodent across a continent in search of its pelt and the riches it will bring. According to naturalist Ernest Thompson Seton (1860–1946), when Europeans first arrived on the shores of North America there were an estimated 60 million to 400 million beavers across the continent. It is an estimate that I often find perplexing due to its broad range, but it has been quoted extensively in both academic and popular literature. What we do know is that some early explorers reported beavers being in every water body that offered suitable habitat. What a sight for a group of people who had already aided in the near-extinction of the beaver's European counterpart back home. In many countries, including the British Isles, *Castor fiber* was already a thing of the past by the 16th century. Overharvesting for food, medicine and fur had left only small pockets of European beavers in the far reaches of Russia and Scandinavia. It would be 500 years before they would once again be reintroduced into the (almost) wilds of Scotland.

So imagine Europeans' delight in finding an Eldorado of fine fur in the waters of the New World. Beaver pelts were one of the great riches of the time, although initially they came second to Atlantic cod. When the explorer John Cabot arrived on what is now Canada's Atlantic coast in 1497, some exchange of furs occurred, but it was a minor event next to the rich cod stocks they sought. By 1532, Jacques Cartier had made a lasting contact with what was likely a Mi'kmaq welcoming party, dressed in beaver furs, along the shores of the St. Lawrence near Chaleur Bay. As the story goes, once all was said and done, the locals went home naked with beads and European tools in hand after literally trading the clothes off their backs.

They left with the promise to bring more furs the next day. And so the fur trade began – sort of.

Cartier's 1542 run up the St. Lawrence River ended at Hochelaga (a.k.a. Beaver Meadows, a.k.a. Montreal), where a set of rapids blocked his progress upstream. This location would later be the site of the great fur fairs of the 17th century. For now, though, the trade in furs was between the European fishermen and the Aboriginal groups of precolonial North America. Cod was a much more attractive commodity and served to feed the palates of Europe. In reality, furs were hardly mentioned during Cartier's voyages and the purchase of them was a mere by-product of coming ashore to dry fish. Back on the European continent, however, felt hats had started to appear by 1456 and then became fashionable across the channel in England by 1510. Just like so many other unexplainable trends, no one could have predicted how a few fancy hats could change the world.

By 1609, the British explorer Henry Hudson was sailing for the Dutch East India Company near what is now the site of Manhattan, along a river that would later bear his name. There, he traded European goods for beaver pelts and otter skins. By 1624, the Dutch East India Company would export 400 beaver skins from the area, and 14,981 pelts by 1635. The trade in beaver pelts was to become a big industry; it was a mammalian gold rush complete with espionage, smuggling, ecological warfare and greed. As the commercial machine of the fur trade – facilitated by middlemen, traders, shareholders and international corporate interests – moved from the Atlantic shore to points west and north, populations of beavers rapidly disappeared across the continental landscape.

The Jesuits reported that beaver were already disappearing from the Three Rivers area of what is now Quebec as early as 1635. By 1638, supplies of European beaver were exhausted and the trade was almost exclusively focused on North America. Ironically, that was the same year that King Charles I of England decreed that only British beavers (a.k.a. those from the Atlantic colonies) could be used in the making of hats and that mixing of other materials into the felt was prohibited. Such a decree solidly confirmed a North American monopoly for British beaver felt. According to Lewis H. Morgan's 1868 book *The American Beaver and His Works*, beavers were already declining in New Amsterdam (now New York City) by 1687, and by 1700 the trade in beavers was reported by Governor Bellomont to be "little or nothing." That they were disappearing in this area likely meant they were declining farther inland as well, where Aboriginal peoples hunted and traded for furs to bring back for the coastal trade. Remember, trade in North American beaver pelts was barely on the radar during the time of Cartier, and populations had been in the tens, if not hundreds, of millions. In about 150 years, beaver had been systematically exploited through a single-minded, corporate effort. And the effort was now to spread west.

Although almost completely associated with North America, the beaver trade was an international phenomenon. The complicated marketing and manufacturing of beaver pelts in Europe and its importance to the British and French economies meant the beaver trade was also implicated in Europe's wars. The "Beaver Wars," fought between England and France and their respective Aboriginal allies, lasted from 1686 to 1713, when, as part of the mediations of related conflicts in Europe, the fighting ended with the signing of the Treaty of Utrecht. Although most aspects of this treaty extended well beyond the fur trade, the results echoed across the Atlantic to the forests and waterways of North America. The signing of the Treaty of Utrecht resulted in France surrendering all rights to the entire drainage basin of Hudson Bay and ceding Newfoundland and Acadia (Nova Scotia) to England. Ultimately, the northern fur areas then held by the French were returned to the Hudson's Bay Company, but dynamic and often cutthroat competition persisted in the shared territories in more southern areas.

The Hudson's Bay Company (HBC), created in 1670, eventually became one of the largest monopolies in the world and covered an area of almost 4 million square kilometres, with even more lands included later. Its evolution into a monopoly was not without setbacks. The French, under La Vérendrye and others, were very successful at trading as far west as the Rocky Mountains, while the British took longer to move beyond their seaside trading posts. With their hold on the trade slipping, the British mobilized to the interior of the continent, where competition with the French pushed both parties farther across what later became Canada. After the fall of Quebec in 1759, the British had an even stronger hold on the beaver trade.

Morrell Allred, in his book *Beaver Behavior: Architect of Fame and Bane*, goes so far as to say "it was the beaver that caused Canada to remain British." As a government-chartered company, the HBC, for almost two centuries, was by almost all accounts one of the unofficial governing bodies of what would become Canada. Its corporate approach to the fur trade in British North America played a significant role in the settlement of this country and the

development of its national identity. As a mer-
cantilist enterprise, where the goal was not only
to expand but to monopolize tariff-protected
industry at the expense of other nations, the fur
trade meant maximum profits for the HBC. For
the Americans, on the other hand, their lands
were settled by agricultural expansion and a
taming of wilderness, rather than the fur trade
monopoly that was developing to the north.
The two approaches established an enduring
difference in the Canadian and American
relationship to the land, something Canadian
economic historian Harold Innis calls the "hos-
tility of beaver and plough."

Although the fur trade was well underway
by the middle of the 17th century, the marketing campaign to recruit more corporate support was just beginning to ramp up. A series
of "beaver maps" were created by some of the
finest map-makers in Europe. I have a replica
of one hanging in my office and show a pic-
ture of it when I give public talks. The image
of zombie-like beavers working their magic
on early Canada catches everyone's attention.

Created by the Dutch cartographer and map publisher Herman Moll in 1715 (and later completed in 1732), it goes by the short title *A new and exact map of the Dominions of the King of Great Britain on ye continent of North America.* The original sold at Christie's auction house for $23,750 in 2008. A similar map made by Italian printmaker Paolo Petrini in 1700 was up for auction at Sotheby's for approximately $40,000 in 2010. In a Postmedia News article, "Rare Map Shows Canada Overrun by Beavers," Randy Boswell wrote an extensive piece about Petrini's map prior to its auction, also describing both Moll's map and another, by Nicolas de Fer, who, like Petrini, based his cartography on a striking image drawn by de Fer's illustrator, Nicolas Guérard. Guérard's drawing was based on descriptions of Canada by a French priest, Louis Hennepin, who visited North America in the late 1670s. In the image, legions of beavers walk on their hind legs while carrying heavy stones on their tails. They look almost robotic in their industrious attempt to work the area surrounding Niagara Falls. There are beavers as far as the eye can see, which to the Europeans meant money and opportunity. Then, as now, maps are the billboards of the world.

These antique maps are relevant even today. At the university, I teach a computer mapping and analysis course called Introduction to Geographic Information Systems. The course is centred on a software package that enables cartography to merge with advanced analytical and database management applications. One of the key concepts I try to teach my students is that map-making is both art and advertising combined, though when I bring Herman Moll's map into the classroom to show excellence in both, I'm never quite sure whether he would be happy or perplexed. Maps must be viewed as representations of the cartographer's bias. Just as one cannot and should not map every blade of grass, the items that are included on a map are the map-maker's choice. What we exclude is either not important or, in the case of some military maps, too important to divulge. We bring forward the items we want to highlight and de-emphasize the things we don't – call it the Mona Lisa smile of the landscape. We also highlight those things we most want to "sell" about the area. Petrini, de Fer and Moll wanted the world (or at least the royalty to whom they presented their maps) to know that hordes of beavers ruled the New World. So many, in fact, that armies of hardworking, well-organized and, most importantly, accessible beavers were eagerly preparing for the arrival of the European market.

It was the ultimate in advertising, and given the length of time it took to finish these maps, they were fine art as well. Moll was even so kind as to describe the Treaty of Utrecht in an insert on his map to emphasize just what wealth rested at the feet of the British. Rule number one of map-making – sell, sell, sell. Rule number two of map-making – location, location, location. It worked.

For many years, a great deal of the southern trade in furs was focused on the Green Bay area of what would become Wisconsin. A journal article by Jeanne Kay, "Native Americans in the Fur Trade and Wildlife Depletion," describes the slow decline of beaver populations in the region. By 1740 eastern Wisconsin was overharvested, and by 1790 beavers were extinct throughout Wisconsin. Kay argues convincingly, however, that in areas that were too remote, had naturally poor productivity or were in constant intertribal conflict, beaver populations likely remained intact. By the 1800s, beaver were in steady decline and many populations were trapped to extinction as the trade moved west. The push was on to obtain as many pelts as humanly possible to feed the markets and economies of Europe.

One of the most ecologically dramatic events of the fur trade era occurred in the Oregon Territory, in the Columbia River watershed, when the HBC launched a "scorched earth policy" shortly after merging with the North-West Company in 1821. The goal was to create a "fur desert," as described by Dietland Müller-Schwarze and Lixing Sun in their comprehensive book *The Beaver: Natural History of a Wetlands Engineer*. Governor George Simpson of the HBC devised a form of ecological warfare at the headwaters of the Missouri River to ensure that American traders were denied access to furs of any sort in that region. As an ecologist, I can only imagine what the complete removal of almost all mammalian life from an entire ecosystem would do to the long-term health of an area. Several studies show the positive effects of beavers on everything from aquatic invertebrates to fish to river otters. The result of this HBC policy was the extirpation of beavers from much of the Pacific Northwest in less than two decades. Gone was a keystone species whose very presence ensured wetland habitats and productive waterways for hundreds of other species.

Just as the physical and behavioural characteristics of the beaver make the species so ecologically important, these same qualities made it perfectly suited to overexploitation. When I conduct annual beaver surveys in the early months of winter each year, it is easy to find their lodges poking above the surface of the ice. Steam rises from the vent hole and a food cache marks the main entrance. The beavers leave an obvious calling card on the landscape. Just as they are easy for me to find, they were easy for the trappers and hunters to find as well. Pursuing a quarry that is non-migratory and colonial was a bonus. If beavers were there in the summer, they would likely be in the same pond throughout the winter, barring extreme drought, disease or predation. One visit to a single lodge could garner several pelts, and during the peak of the fur trade, beaver were more valuable than all others, such as wolverine and lynx, which later became known as "fine furs." It was the underlying felt of the beaver's fur that was so valuable in the manufacture of hats.

As Harold Innis's classic book *The Fur Trade in Canada* outlines in remarkable detail, the pelt enterprise in North America was not just a romantic exploration of the New World. According to Innis, it created a made-in-the-wilderness economy that set the stage for resource extraction and exploitation for centuries to come. How it affected the Canadian ecological psyche is anybody's guess.

Entire groups of indigenous people readily abandoned their stone tools, and at times their agricultural lifestyles, to pursue an animal that previously held no elevated role in many (but not all) Aboriginal cultures, likely due to the beaver's ubiquity. Although many aspects of their older technology persisted, Aboriginal groups were heavily involved in trading furs for European staples. Some authors argue that, prior to the fur trade, white-tailed deer and caribou were the most valued species for many of these groups, although the Beaver Bundle, connecting its owner to the power and spirits of the Underwater People, was a very old and important aspect of the Blackfoot culture in the west. In eastern North America, beavers were worn as clothing and eaten, but really were inconsequential to most of the Aboriginal people at that time. It must have seemed remarkable to receive such rare items as the Europeans were providing in exchange for just some worn pelts.

Many historians note that it was a mutually beneficial arrangement for all involved.

European trade goods were highly sought after by the Aboriginals, and access to furs and means of travel were desperately required by the Europeans. To paddle a canoe well meant access to the heart of the fur country and a goldmine of pelts. Various Aboriginal groups who had not previously met were now in competition with each other or working with other bands who acted as middlemen for either the French or the English traders. Dr. Patricia McCormack's wonderful *Fort Chipewyan and the Shaping of Canadian History 1788–1920s*, is a testament to the complexity of these relationships and the effect of the fur trade on previous ways of life and changes in "modes of production."

It was not just the ecological world that was turned upside down. Those Aboriginal groups who could manage the corporate takeover of a continent were the ones who found success in a rapidly changing world. Innis includes a quote from Nicolas Denys's 1672 writings that describes how the Aboriginal peoples residing around the Gulf of St. Lawrence had already abandoned their own utensils and replaced them with European goods because of the fur trade. Although people argue it was the introduction of the steel trap in the 1800s that accelerated the demise of the beaver, many areas had been depleted long before then. According to zoologists Stephen Jenkins and Peter Busher, by 1900 beavers were almost extinct in North America and in some areas they have yet to return.

By 1778 the fur trade reached the Athabasca and Mackenzie regions in northern Canada, where the quality of the fur made all else seem substandard. Fort Chipewyan was the hub and became Alberta's first town, established in 1788. An epicentre of the struggles between the North-West Company, the XY Company and the HBC, it was the site of hostage-takings, murders, fierce rivalries and finally the merger of these companies (or remnants of them) into the HBC. Controlling the access to the Mackenzie River, the rich Peace–Athabasca delta and the area that would later become Wood Buffalo National Park was key to dominating what had now become an unstoppable commercial machine.

The distance from the key ports of Montreal and the deep-water ports along Hudson Bay made the import of trade goods and the export of beaver pelts expensive. To be profitable, the business needed to run like a well-oiled machine, so efficient modes of transportation and reduced labour needs were critical. Traders and voyageurs were central to the operation, collecting and bringing the furs back to the eastern forts for transport back to Britain. To be an HBC trader or a voyageur was not easy and these postings were often very challenging. However, working for the company was still better than most opportunities presented back home in Britain. The life of independent traders, including the coureurs de bois, offered more freedom. Still, the romantic life reflected, for example, in Frances Anne Hopkins's famous 1869 painting *Canoes in a Fog, Lake Superior* was more myth than reality. The reality was all about money, and the more beaver pelts that arrived at the fur trade posts of the HBC, the better.

According to HBC records, over 4.7 million beaver pelts arrived at the auction houses of Britain between 1769 and 1868. These figures do not include the great numbers of pelts that did not make the grade and were discarded. This period was long after the height of the trade, which had started to diminish, but not disappear, after the 1820s with the growing popularity of silk hats and the corresponding decline of beaver felt. That beaver were becoming more difficult to find did not help either. In 1869, the fur trade post at Fort Chipewyan recorded 24,679 beaver pelts brought in from the Athabasca region, and again those were just the ones deemed acceptable. When looking at the records, it is difficult to comprehend the vast numbers of furs actually harvested. It is even more difficult to assess the ecological implications of the loss of millions of beavers across an entire continent.

There were moments of reprieve for the beaver when various Aboriginal groups were at war and thus not out hunting and trapping beavers for the French and English. These were frustrating times for the merchants, who wanted a stable and constant flow of thick pelts to their warehouses. There were also diseases, such as small pox and influenza, that wiped out many of the people who were so essential to the enterprise. Whereas much of the trade in the United States was conducted by white trappers and controlled by powerful business-men like John Jacob Astor, who made his fortune from the fur trade, the Canadian trade was completely dependent on the partnerships with indigenous peoples and, in later years, on the monopoly of the HBC. Although white set-tlers eventually played a more important role in the Canadian fur trade as they developed agri-cultural land across the prairies and trapped on the side, the wealth of the HBC's governors and investors was built with the help of Aboriginal peoples all across Canada. Although the mo-nopoly of the HBC ended in 1870, the fur trade continued into a new era of competition that reached new heights in the 1920s before level-ling out by the Second World War. Today there are still echoes of a modest trade in furs, but the days of the coureur de bois and voyageur are very much a thing of the past.

It was through the waterways that Canada was opened up to a natural resource economy. In his book *Fur Trade Canoe Routes of Canada: Then and Now*, Eric Wilton Morse provides an excellent account of how the pursuit of beaver allowed access to even the most remote areas of what was to become Canada. At profits of 1000 to 2000 per cent during the most successful years of the trade, it was a lucrative means to claim a country. It is amazing that the beaver survived at all. For during this time, the passenger pigeon, the sea mink and Dawson's caribou were not so fortunate and quickly became extinct. From trading at the river mouth at the beginning, to the later cross-Canada expeditions by HBC governor George Simpson and explorer David Thompson, the stories of Canada unfolded like those of no other nation in the world. The country is steeped in fur trade traditions, that created a mystique of the land and a myth of superabundance on which we still base our "wilderness economy." In many ways Canada is a country with a split personality, one that defines itself by the very wilderness it nearly destroyed.

The Dry Years

Although beavers managed to survive three hundred years of commercial exploitation in North America, by the end of the "early" fur trade in 1870 their numbers were marginal at best. As early as 1883 beavers were legally protected in the territory that would later become Alberta, and by 1907 the new province had passed The Game Act, imposing a five-year moratorium on harvesting beavers. Even prior to government intervention, the Hudson's Bay Company had attempted to impose some conservation measures in the 19th century. In 1913 limited trapping was allowed by the Alberta government, and increased harvests occurred as populations slowly recovered. In 1936 trafficking in beaver skins was placed under federal control in the United States, and in 1938 beaver trapping was closed entirely in Ontario. Small comfort when many areas had no beavers left to trap in the first place. Earlier laws to protect resources were largely ignored and poorly enforced. In many places, beavers had been extirpated by the mid-1800s and would be gone well into the mid-20th century. Given their ability to create and maintain wetlands, the loss of this keystone species would have had devastating effects on water resources throughout North America.

It was no longer just the fur trade that compromised beaver populations; the creation of agricultural land and colonial settlements resulted in the unprecedented draining of wetlands and diversion of rivers, streams and creeks. In the United States, draining wetlands was already common by the mid-1700s. In their article "History of Wetlands in the Conterminous United States," Thomas Dahl and Gregory Allord noted areas of significant wetland loss between 1800 and 1860 in most of the states east of a line extending from Minnesota in the north to Texas in the south. In another paper, Dahl reported that by 1990, six states had lost at least 85 per cent of their wetlands and twenty-two others had lost over 50 per cent.

The situation in Canada was no different. Natural Resources Canada has identified the main factor in wetland loss to be agricultural activities, which account for over 85 per cent of wetlands lost since European colonization. Large expanses of Ontario and the Prairies have lost over 70 per cent of their original wetlands, while 65 to 80 per cent of our coastal marshes have also been destroyed. A collaborative report by the federal, provincial and territorial governments, "Canadian Biodiversity: Ecosystem Status and Trends 2010," indicates that up to 98 per cent of the wetlands near Canada's urban centres have either been lost or degraded. For semi-aquatic species, such as the beaver, making a comeback without adequate habitat is almost as challenging as outrunning the fashion industry.

With both massive declines in surface water due to wetland loss and the killing of millions of beavers that would otherwise help establish such wetlands, it is curious that those who develop models for both historical and future droughts rarely incorporate these variables into their calculations. No doubt El Niño and La Niña years play an important role in climatic trends. Ocean currents are critical for their hydrological inputs into the global water cycle. But would not the loss of at least 70 per cent of the wetlands on the third largest continent in the world, and the near extinction of a furry rodent with a penchant for engineering waterworks on not one but two continents, have some effect? For local weather systems, water bodies play an important role in evaporation and localized cloud formation. What goes up must come down. But what if there is nothing left to go up?

Drought is one consequence. For those who wish to search out an authoritative definition of "drought," it will be a disappointing venture. There are several types of drought: meteorological, agricultural, hydrological, even socio-economic. What they all have in common, however, is a lack of moisture where it is needed most. With meteorological drought there is a distinct lack of precipitation. Agricultural drought relates to both inadequate precipitation and lack of soil moisture that would support crop production. Surface water is important and groundwater is critical. With hydrological drought there is reduced supply in water bodies such as streams and underground aquifers. As for socio-economic drought, some might think it should mean better role models on reality TV, but it relates to the social and economic effects that develop when water supplies dramatically decline or disappear.

One aspect of my PhD research was to quantify the effects that beavers had on the extent of open water across the landscape during times of varying climatic conditions. Although I couldn't possibly have planned it that way, my research happened to fall within the worst drought on the Canadian Prairies in 137 years. For most wetland ecologists, the loss of many of my research sites would have been a disaster, but it turned into the biggest gift one could ask for. I was able to see how beavers adapted to extreme drought and then could compare these effects over time.

In the short term, it was obvious that dry conditions were just another test of the tenacity of beavers. As the drought progressed, they dug channels throughout the pond bottom to collect any and all available water. Much like the settlers who used drainage channels to de-water wetlands to dry their fields and create agricultural land, beavers excavated channels to maintain and direct water where they needed it most – at the entrances to their lodges and along access routes to their favoured foraging areas. In some cases they spent more time digging channels with their two front paws than collecting food for their winter food caches. The results were amazing. By the time the drought of 2002 was in full swing, the ponds that still contained water were most often those occupied by beavers. Although some families of beavers packed up and waddled down the road in search of wetter pastures, most stayed put and successfully survived a winter in their lodges and under the ice.

Although difficult to compare worse with worst, cattle ranchers were one of the agricultural groups hit hardest by the drought in the region, and it was common for them to request access to a neighbour's property if there was an active beaver lodge on it. So severe was the drought that those ponds were some of the only places that still had water. Not only did cattle get access to water, they could also forage on some of the only green vegetation left in the area.

The drought was not without consequences for beavers, though. An estimated 7 to 10 per cent of Elk Island National Park's active colonies did not survive the winter following the drought. For a semi-aquatic animal, however, that is a remarkably low number. In ponds that were too shallow and froze to the bottom during winter, the beavers chewed their way through the sides of their lodges to escape and find food. In these instances, coyotes would happily wait at the sound of chewing, until beavers emerged and provided a handy meal. Finding the remains of a coyote feast was not uncommon beside some of the lodges. More recently I have seen beavers react to more prolonged droughts and have realized that much more than 10 per cent of the population can be lost. Nevertheless, the same observation holds true: where there are active beaver colonies, there is water. Ponds without beavers usually dry up first and are the last to refill during the years following a drought.

Apart from my immediate observations of the ability of beavers to survive drought and keep water on the landscape, my doctoral adviser, Dr. Suzanne Bayley, and I combined three long-term data sets to assess how beavers affect the extent of open water during various climatic conditions, including drought. In our paper, "Beaver (*Castor canadensis*) mitigate the effects of climate on the area of open water in boreal wetlands in western Canada," we compared fifty-four years (1948–2002) of beaver-lodge occupancy data, precipitation records and the extent of open water in historical aerial photographs and discovered that beavers had a significant effect on the increased resilience of a landscape to even extreme periods of drought. Using data from twelve of those years, for which we had complete photographic coverage plus beaver census and climate data, we found that beavers accounted for 85 per cent of the open-water area regardless of the amount of precipitation or extremes in temperature. We were fortunate to be able to examine sites that had been without beavers for over a hundred years and those same locations once beavers had become well established. By looking at seventy wetlands prior to the reintroduction of beavers and those same wetlands with and without active beaver colonies over time, we determined that ponds with beavers had nine times more open water area than those same ponds without beavers, even during times of drought.

But our most striking finding came when we compared two periods of drought: 1950 and 2002. Although 1950 was the fourth-driest year on record, it still had 47 per cent more precipitation than 2002, the driest year on record. However, in 1950 there were still no beaver present in our study area, and as a result there was 61 per cent less open water than in 2002, when beavers were well established. In other words, beavers were actually mitigating the effects of drought.

What's more, some of my more recent research shows not only an increase in the surface area of open water, but also that beaver ponds tend to be deeper than nearby ponds because of the presence of beaver channels along the pond bottom and excavations near the entrances to the lodge where the food cache is constructed. Given that most of my study sites are on isolated ponds, which at best might be connected to an intermittent stream, this speaks to the ability of beavers to maximize local water capture without much influence from extensive stream or river systems.

Although Suzanne Bayley and I used aerial photographs dating back to 1948 for our research, there exist earlier images of the terrain as well. These too proved quite revealing, not only as to beavers and drought but also to illustrate the limitations of aerial photography as a mapping technique. Pictures taken from an aircraft can be either vertical (directly over the landscape) or oblique (at an angle to the landscape). With vertical photographs, such as the ones used in our study, the ground appears as if one were suspended directly above it, like a bird. This perspective allows the viewer to calculate the exact extent of forest cover, water or similar surface features with relative ease. Oblique photos provide more depth and definition, but they are difficult to use when attempting to calculate areas accurately. I was able to locate a couple of sets of photos from the 1920s that showed several parts of my study terrain. Most of these views were oblique, however, and there were not enough vertical ones to provide complete coverage. Still, what I could see from them was fascinating. The early 1920s had almost twice the precipitation of 2002, yet the landscape nevertheless looked parched. Ponds that had water in 2002 were completely dry in the 1920s. The contrast was so stark that it was as if a bomb had dropped in some places and evaporated all the water off the landscape. A world without beavers was most evidently a world without water.

Other researchers have also found remarkable impacts of beavers on hydrology. During her doctoral studies in Colorado, Cherie Westbrook determined that the presence of beaver dams on a river resulted in an elevated water table even during the drier summer months. Most importantly, the presence of a beaver pond positively affected groundwater storage both adjacent to and downstream of the pond. These effects were equivalent to at least a one-in-two-hundred-year flood event. Not only did the ponds create a larger extent of wetted soil, the soil stayed wetter for longer than during regular flood events.

Still other studies have determined that the presence of beavers was the major factor behind the quantity and type of wetlands present, and that beaver activity continues to affect ecosystems long after the beavers have left.

With such a dramatic influence on hydrology at a landscape level, is it possible that the impacts of the forty long-duration droughts that have occurred in western Canada over the past two centuries could have been reduced by the presence of beavers? We may never know whether the Dirty Thirties throughout North America or the "Civil War drought" of the 1850s to mid-1860s in the United States could have been less severe had there been more beavers present and fewer wetlands destroyed. Richard Seager and Celine Herweijer's paper "Causes and Consequences of Nineteenth Century Droughts in North America" targets the Civil War drought, along with overhunting, as a major factor in the near-extinction of the American bison. Could beaver have helped save one of the largest land mammals on earth? Who knows? But what we do know is that beaver can mitigate the devastating effects of drought and that a combination of wetland loss and the extirpation of beaver in many areas by the mid-1800s had to play some role in the recovery of the land and its wildlife from these drier periods in our recent history. The beaver is a species that has evolved through many major climatic shifts. Drought is just one of the many challenges it has faced over the millennia.

Return Crate

Ironically, beavers had been eliminated from the heart of the Beaver Hills of east-central Alberta by the mid-1800s. In 1913, some of these lands became Elk Island National Park and benefited from protected status, but it would be almost thirty more years before beavers would swim there again.

With the HBC's Fort Edmonton less than 90 kilometres away, the region was a hub of fur trade activity for trappers in the North Saskatchewan watershed. In reality, there were five "Fort Edmonton" locations from 1795 to 1891 as the site moved around central Alberta until the final trading post was established at the spot where the Alberta legislature building is today. Regardless of the mobility of the forts, their effect on beaver in the region was devastating. Not only were the beavers completely extirpated but forest fires, agriculture and drought changed the landscape for years to come.

In what some might see as a visionary move in the 1940s, Elk Island's park superintendent decided to bring beaver back to help restore the former ecosystem. Seven beavers from Banff National Park were successfully introduced into Elk Island's Astotin Lake in September 1941. For several years I worked as a warden in the park, and late one night while on a standby shift I came across the original letter and then telegram from the superintendent of Banff National Park documenting the arrangement. The letter read:

> June 30, 1940
> Dear Sir:
> re: Beaver.
>
> By letter of October 2, 1940, you reported that three or four pairs of beaver for Elk Island National Park could be captured

for approximately $75.00. They are to be taken by truck to Elk Island Park.

It would be appreciated if you would make arrangements to carry out this work during August of this year.

Yours faithfully,
J. Smart,
Acting Controller.
The Superintendent;
Banff National Park, Banff, Alberta.

Once the deal between the two park super-intendents was made, the telegram (and beavers) followed. As with the letter, the telegram was short and to the point. Dated September 12, 1941, it read: "BEAVER SHIPPED C.P. EXPRESS TRAIN NUMBER FOUR TODAY PLEASE MEET TRAIN STOP RETURN CRATE." Apparently beavers and crates were of equivalent value, although I did notice there was not much fanfare about parting with the beavers.

This reintroduction of beavers represents just one of at least two such initiatives in Canada's national parks. In 1948, beavers were flown from Prince Albert National Park in northern Saskatchewan to Wood Buffalo National Park, which straddles the border between northern Alberta and the Northwest Territories. The park is also immediately adjacent to the former heart of the fur trade, Fort Chipewyan. As difficult as it is to believe, Canada's largest national park, at 44,807 square kilometres, had run out of beavers, and it was this fresh supply from Saskatchewan that allowed the population to recover. All across Europe, Canada and the United States there was a movement afoot to bring beavers back to the landscapes where they had once lived.

Many authors credit Grey Owl (Archibald Belaney) for inspiring the beaver reintroduction and conservation programs that became popular following the drought of the 1930s. As a writer and wilderness celebrity, he ran his successful campaign to save beavers and the wild places they represented from his cabin homes, first in Riding Mountain National Park and then in Prince Albert National

Park, where he worked as a naturalist for the Dominion Parks Service (Parks Canada). In many ways he and his Mohawk wife, Anahareo (Gertrude Bernard), were Canada's first famous conservationists, and their two little beaver kits, Jellyroll and Rawhide, were the stars of the show.

The better known of Grey Owl's two cabins rests on the shores of Ajawaan Lake in Prince Albert National Park, deep in the heart of the boreal forest. In the 1930s, when the famous couple lived in the park, hundreds of people made the pilgrimage to visit Grey Owl and his "beaver colony" at his remote cabin. The most curious thing is not that the couple lived with two beavers that had built and lived in a lodge inside the cabin, but rather that these pilgrim tourists actually believed that this tall Englishman really was the son of a Scottish father and Apache mother who had been part of the Wild Bill Hickok Wild West Show. It was not until after his death at the age of 49 in 1938 that the truth was fully revealed. It was a shock to Grey Owl followers throughout the world, but regardless of his deception, his conservation efforts had long-lasting effects.

To journey to Grey Owl's cabin is to venture into recovery of spirit and environment. The cabin feels almost as remote now as it must have felt for those hiking there during the 1930s. The last time I made the trip, I took along a good friend who had never been there. Although I had often heard the story of Grey Owl as a child in the 1970s, the story, as well as the area, was new to her. Being friends with someone who has studied beavers for so long has definitely increased her success when playing the Canadian version of Trivial Pursuit.

We paddled my canoe to our campsite just as the sun was setting. By the time the tent was up and we had settled in for the night, a loud hoot of a great grey owl rang through the darkness. We were definitely in Grey Owl country. Despite the length of the paddle, the battle with insects and the unpredictable weather, the trip was a worthwhile effort, and when we arrived at the cabin the next day, my friend was awestruck by its remoteness and the fact that people travelled so far in the 1930s just to see this man, especially given his controversial character. Of course, all that now remains of Grey Owl, Anahareo and their daughter, Shirley Dawn, are three headstones upslope of his cabin, which is just a short walk from Anahareo's. I am told Anahareo built her cabin on a small hill above Grey Owl's to escape the sound of beavers chewing all night and the crowds of tourists that swarmed his cabin all summer. The beaver lodge still sits inside the cabin at the lakeside edge. On one of the walls is Grey Owl's paddle. The cabin serves as a symbol of his love of wilderness and belief in beavers as a restorative feature of the ecosystem.

Although Grey Owl was likely the most famous and charismatic champion of beavers in the early part of the 20th century, a less-known but equally visionary Brit was working to bring beavers back into the heart of British Columbia. Eric Collier wrote a wonderful book about his life with his wife and son in the heart of the Meldrum Creek valley, near Williams Lake, British Columbia. His book *Three*

Against the Wilderness details their life together on a remote trapline over a thirty-year period, from the early 1930s to 1960. When they first moved to the area, streams were dry, wildlife was depleted and forest fires were a constant threat. Overharvesting of beavers and other wildlife had taken its toll on the land. Before moving onto their trapline, Collier and his wife, Lillian, promised to fulfill her First Nations grandmother's request to bring beavers back to the land so that it could be as it was during LaLa's childhood, before the White Man came.

In 1932 the Colliers began working tirelessly to rebuild the dams that had fallen to ruin after the extirpation of beavers from the area. For the first time since the demise of the beavers, the land began to come back and wildlife moved in once again. The Colliers' efforts also resulted in critical flood control for areas downstream during spring snowmelt and allowed for more consistent water supplies for the entire region throughout the summer. Their efforts did not go unnoticed. In 1941 a conservation officer drove to the Colliers' cabin on a very rough road, dropped off two pairs of beavers and promptly drove away. By 1950, the beaver population was healthy and the Colliers were able to trap a modest number. The entire valley was vibrant with wildlife and wetlands.

Although Collier's book is rarely listed as a must read on a Canadian bestseller list, it is still in print and it is a jewel that speaks not only to the capacity of beavers to restore the environment but also to the passion in those who champion them. Ironically, I still receive phone calls regularly from people who ask, "Has anyone ever thought of reintroducing beavers to help restore aquatic habitats?" With a quick look at my bookshelf I smile and think of a quiet, honest man named Eric Collier and his wife, Lillian, who devoted their lives to doing just that.

Reintroducing beavers into their former habitat really was a global phenomenon. The essays in Busher and Dzięciołowski's anthology *Beaver Protection, Management and Utilization in Europe and North America* provide an extensive summary of beaver extirpations and reintroductions on both continents, but especially in Europe. After being extirpated in Finland and Sweden by the second half of the 19th century, European beaver were reintroduced from Norway between 1922 and 1940. In Denmark, beavers had been hunted to extinction during the Bronze Age (3000 to 600 BCE) but were then reintroduced in the 20th century. Similarly, in the Baltic region, which also had seen significant hunting as early as the third millennium BCE, reintroduction programs were begun in the early to mid-1900s.

There has been a second wave of reintroductions since, mostly in Britain and Europe, although there is also a bit of a revival in some areas of the United States. In Austria, reintroduction programs began in the 1970s and 1980s. Bratislava also began reintroducing beavers in the 1980s. After being hunted to extinction in the 1800s, European beavers were reintroduced into the Netherlands in 1988 and 1991. The most recent reintroduction brought European beavers back to Scotland at the end of May

2009 after a 400-year absence. Reintroduction programs are not without controversy, though. Beavers are beavers and settled areas are settled – complete with roads to flood, trees to fell and conflicts to occur. In some areas of Europe, beaver reintroductions have been so successful that conflict management is the new phase of ecological restoration.

While attending the International Beaver Symposium in 2006, my colleagues and I toured a rural agricultural area where fish ponds were infiltrated by beaver burrows, beet fields were candy to the little rodents (much to the frustration of the farmers), and live-trapping and relocating the animals to new areas was the norm. By the time the next beaver symposium was held three years later, it was becoming increasingly difficult for European wildlife officials to find homes for the furry refugees. Once again it comes down to habitat. Just as wetlands have been drained and the land "reclaimed" for agriculture in North America, so have cities, resorts and a burgeoning human population filled much of the habitat that once belonged to beavers in Europe. Now there are beavers swimming down the canals of Berlin and sharing the sights along the entire extent of the Danube. To the dismay of adjacent land-owners, they are also chewing, building and excavating their way into both built and natural landscapes. Regardless of potential conflicts with humans, however, reintroductions of beavers have been used to restore ecologically damaged landscapes on both continents.

In the interior Columbia River basin of Oregon, which was ecologically devastated by the HBC's bare-earth policy in the 1820s, the presence of beaver dams has proven to be the perfect solution for rehabilitating unnaturally incised streams that had been subject to constant erosion. In their paper "Geomorphic Changes Upstream of Beaver Dams in Bridge Creek, an Incised Stream Channel in the Interior Columbia River Basin, Eastern Oregon," Michael Pollock and his associates examined several eroded channels leading into the river that had recently been blocked by beaver dams. These were channels that had formed from human disturbance and subsequent soil erosion. The dams allowed sediment to build up behind them, which rapidly improved riparian habitat and restored the stream to a more natural and stable condition. Other organizations in both British Columbia and the Pacific Northwest of the United States are also looking to beaver to help restore damaged salmon habitat and aid failing fish stocks. Once again, putting the proverbial keystone back into the arch can benefit so many other species.

When Suzanne Bayley and I published our paper on beavers and drought, she warned me it was going to receive a lot of attention. Having worked so long for the federal government as a warden with Parks Canada, I was used to being a fairly low-key civil servant. The paper came out shortly after I had accepted a faculty position with the University of Alberta, so I was much easier to find, especially on the Internet. The paper drew international attention. As soon as I would hang up from one phone call from a man asking if I had any spare beavers I could release onto his property, the phone would ring again with someone seeking advice on how to get rid of the animals. If I had been smart, I would have set up an Internet dating service to link up those seeking rodents with those wishing to dispose of their overzealous ecosystem engineers. I could have made millions. More than two years later, the phone calls, speaking requests and pleas for wildlife management advice continue. I have my regulars, whom I adore. They call every few months just to discuss the latest wetland issues or wildlife rescue stories.

Having worked both sides of the management issue, I can understand where both groups are coming from. It is frustrating to have expensive facilities, trees and equipment damaged, but it is also important to appreciate all the ecological benefits that are part of a beaver's world. In many situations, it is also time to stop seeing the beaver as a pest and more as a pilot to navigate us through a warming climate and the complex ecological challenges it brings.

When I presented our "drought paper" to an international conference, one person approached me and said how thrilled he was that my paper conclusively confirmed his position that beaver reintroductions were absolutely necessary and should not be delayed. Shortly after he left, a land manager came up and enthusiastically thanked me for my presentation because it proved without a doubt that swift and decisive management was absolutely necessary to cope with the beaver problem. Both left happy and vindicated with what they heard during my talk. I left with a curious smile on my face, knowing that somehow our research had sounded a very deep chord.

The value of the original beaver reintroduced into Alberta's Beaver Hills was $75, as long as the crate was returned. The benefits of beaver activity to farmers, wildlife and ecological systems during the drought of 2002 were immeasurable. This dichotomy between beaver as hero and beaver as pest has been a mainstay in the North American psyche for centuries. It seems the only time it was unanimously regarded as a valuable and priceless gem was when it helped open up a continent to feed the coffers of Europe.

Conflict and Creativity

Although live-trapping and relocating beavers to reintroduce them into their previous habitats are still management options in some areas, this situation is far from the norm. Throughout my career with Parks Canada, dynamite, backhoes, shovels, culvert guards, firearms and all other measures of creativity were much more common. Live-trapping and relocation did and still does occur, but it is not without risks. Relocating wildlife always has potential for conflict with resident animals, poor habitat provisions or the introduction of disease carried by translocated animals. Other creative methods have also been employed. For example, modifications to culvert designs help alleviate flooding in some areas. But creativity often takes two things: time and willingness to change.

The more expedient management methods are not unique to Parks Canada; they are ubiquitous in any wildlife and land management agency. Trying out new techniques can be costly, initially time-consuming, and risky. And let's face it, using dynamite can be fun. However, I have seen some beaver dams that were loaded with dynamite remain relatively intact after the explosion, which only resulted in a mere trickle of water running over a slightly damaged edge of the dam. For years I heard the comment "this is the only thing that works," when nothing new had been tried for twenty-five years. But at the same time, I also had some of the least likely people say, "They were here first; we need to learn to live with beavers and stop building things next to their dams and in the floodplains." Innovation often begins when you least expect it.

I think one of my most transformative moments was when, as a park warden, I was told to shoot out a colony of beavers that had flooded a culvert and damaged a road. Whether my first couple of shots were effective or not, I will never know, but I do know that I got back into my warden truck, drove to the office and said there was a better way to manage the problem at that particular location. Having seen forty years of beaver occupancy data for that exact pond, I knew that no matter how many beavers we shot (and the colony had already been removed several times over the past decades), a new family of beavers would return within a few short months. Installing a flow device and monitoring its effectiveness would have been a much more cost-effective and ecologically sensible thing to do. The pond, after all, was also filled with amphibians and nesting waterfowl at the time.

I argued for taking an adaptive management approach, one that would treat management actions as an experiment where we could assess our success and rework our approach as things progressed. This form of management is based on the precautionary principle, which in short translates to "better safe than sorry." Adaptive management is a progressive science and is not an entirely new concept; it has been in the ecology textbooks since the late 1980s. The response I received was a resounding "my way or the highway." For an organization that prided itself on using the terms "adaptive management" and "ecological integrity" in its guiding principles and policies, this attitude left me dumbfounded, especially after almost twenty-five years experience working in protected areas.

Many other national parks where I had worked were doing just what I had suggested, but adaptation is not everyone's evolutionary strength I suppose. I never did go back and shoot the beavers, but they were shot by morning all the same. The pond was duly drained and the problem solved until the next batch of beavers arrived shortly after. The road still floods, more beavers are shot, the pond is still subject to draining by dynamite. And since it is good habitat, beavers will return again and again.

Our environmental legislation is no different, really. In the province of Alberta, the Water Act states that a statutory approval is required before draining or infilling a water body, and the Interim Wetland Policy strives for "no net loss" of wetlands. However, under the Water Act, a beaver dam can be removed (which would result in draining the water body) without an approval as long as the person owns the land or lives adjacent to the land containing the pond. Conversely, to protect wildlife habitat provided by wetlands, Alberta's Wildlife Act makes it illegal to remove a beaver dam unless one obtains a permit under the Agricultural Pests Act, the Water Act or the regulations under the Wildlife Act. The Wildlife Act also prevents the harm or removal of beaver houses (lodges or dens) on public land. Other provincial and federal policy and legislation is almost as confusing, and actions within protected areas often remain as inconsistent as those outside their boundaries. My experience with the persistent beaver colony mentioned above is likely not an unusual one.

There is the undeniable reality, however, that people are moving farther into what is left of not only beaver habitat but the remaining habitat of several other species of wildlife. At Christmas, illuminated plastic deer decorate lawns throughout rural residential "estates," perhaps as a form of reverence and appreciation for the wildlife habitat that was once there. All is well and good until the first tree is felled, the favourite plants are eaten from the garden or the driveway is flooded. We want to co-exist with wildlife, but we want it to be well-behaved wildlife, much like our lawn ornaments. As is the case with children, they are cute until they find a crayon and a wall.

The industrial-scale fur trade is truly a thing of the past. Although there are still wilderness trappers and the Northern Store, the bottom has all but fallen out of the fur market. When I visit with elders and trappers in remote communities, they talk at length about their relationship with the land and how so many of their children and grandchildren have no desire to go into the bush. Why would they when they can make more than a trapper's annual income in just two months in Alberta's oil sands? But even at eighty-five years old, some of them still insist on going out to their trapping cabins regularly to maintain their traditions and sense of place. It is their interactions with the land and its rhythms that allow us the most remarkable insights that a three-year research study could never reveal.

Today, synthetic fabrics have largely replaced the products of the silkworm and beaver of the past. If furs are worn, they are what we call "fine furs," from animals such as marten, fisher, wolverine and lynx. We obtain many of our clothes from the petrochemical industry now. Their factories have a whole new impact on the environment that could never have been anticipated during the fur trade.

But there is a new need for trapping now, thanks to urbanization and the human population explosion. One of my acquaintances, a full-time trapper, explained to me that most of his trapping revenue does not come from the bush trapline he has tended for at least two decades. It comes from removing "problem wildlife" from cities like Calgary and Edmonton. During the dark of night he is the one who removes the culprits who cut down Mr. and Mrs. Urban's favourite ornamental cherry tree, flood the riverside running trails or attack dogs that jump into the pond to chase balls and sticks thrown right next to young beaver kits swimming beside their lodge. Cities and counties receive the complaints and call the trappers, and like the police the trappers take away the perpetrators. In the morning, the complainants can enjoy nature again in peace.

My acquaintance once told me he would be fine with not trapping in cities. After all, concerned citizens had called the police several times to report "a strange man" sitting by the side of the river during the wee hours. In response to these concerns, he said he would be happy to bring in five or more black bears and a pack of wolves into downtown Edmonton (a city with a population of just under 750,000 and a metropolitan population of over a million) to allow natural predators to keep beavers under control just as they would in a more natural environment. Whether the citizens of any urban centre would agree to such a move is doubtful. Modern *Homo sapiens urbanus* has forgotten that nature has a balance and that, with the loss of all our large predators from much of the Prairies and Great Plains of North America due to overhunting and extensive habitat loss, human predators are one of the few choices left to control wildlife populations. The coyote, a medium-sized predator that has exploded in population since the loss of its larger cousin the wolf, is much more content to take the odd house cat and eat mice and other city critters as it strolls through our urban green spaces at night. Beavers can be a hard catch with their big teeth and aquatic escape routes.

It would be nice if we could relocate beavers when problems arise, and trappers sometimes do, but there is so little unoccupied habitat left, for any species of wildlife, that the options can be limited. Being fiercely territorial animals, putting an unrelated beaver into another family's habitat would result in a fight at the very least and expulsion at best. There are other solutions for beaver/human conflicts, but, again, they take time and patience that are so rare in today's instant-everything world. Just as we charged through millions of beaver pelts during the fur trade era, we are now expanding our industry and urban areas into the remaining forests and unprotected wilderness. Less habitat is available and beavers searching for a suitable pond site can easily come into conflict with humans.

There are a few animal rescue groups who chat with me from time to time, and so often they are dealing with orphaned beaver kits whose parents were killed during the course of urban development or by a trapper or by someone's dog. These volunteers raise these beavers by hand and search for the increasingly rare site that could accommodate their wild release without conflict with other animals. Raising a beaver is not an easy task. Just ask Audrey Tournay, author of *Beaver Tales: Audrey Tournay and the Aspen Valley Beavers*. She is a woman with a heart of gold who has dedicated her life to the Aspen Valley Wildlife Sanctuary in the Muskoka region of Ontario, which she established and helps run. For over thirty years she has raised orphaned beavers in her house, in the special ponds constructed in her yard, and at the sanctuary. During one phone call, I heard the distinct sound of beaver vocalizations in the background and when asked, she replied, "Oh, the beavers want to say hello." An orphaned baby beaver had crawled up on her shoulder and was nosing up to the phone to give its distinct nasalized greeting across the airwaves. Tournay's book is a lovely journey through the activities that only beavers could create inside a wooden house with wooden furniture and wooden doors. Having raised a puppy or two, I am not sure how she has managed through dozens of beavers, kits and adults alike. Her house would be the equivalent of a pair of favourite slippers to a new puppy.

Hearing both sides of the story – the trappers' and the wildlife rescuers' – I am often left wondering who the true culprit is. So many times I hear the "two sides" describe each other in less than endearing terms, but somehow I think it is the average Canadian who gets off the hook too easily.

There is a flaw in the Canadian identity. Watching the closing ceremonies at the 2010 Winter Olympics in Vancouver, millions of people gazed at beavers and moose dominating the main stage as if they were as much a part of being Canadian as winter itself. A quick walk through any store in Jasper or Banff national park would find you surrounded by wildlife motifs on everything from underwear to cribbage boards. Wildlife sells and, as such, it is the perfect image of Canada. While working on my PhD, I received every type of beaver gift imaginable: beaver slippers, beaver notepads, beaver ball caps, a Playmobil beaver, beaver hand soap, beaver droppings (a form of chocolate candy), beaver stuffed animals, beaver T-shirts, beaver energy drinks, beaver raku pottery, beaver badges and the list goes on. Ironically, none of these products had even a trace of actual beaver in them; most were made in the factories of China. Years later, I am still receiving a steady flow of beaver gifts. Good thing I chose not to study skunks.

In North America we have always had a "myth of superabundance," a term coined by Stewart Udall in his bestselling classic *The Quiet Crisis*. The phrase alludes to the belief that we have an unlimited supply of natural spaces and resources. It is a belief that was reflected in the cod fishery, the harvesting of our forests, the exploration of the continent in search of furs and gold, and now our exploitation of our oil, gas and mineral resources. Next it will be water. If we consume it all, more will come somehow; that is our irrational belief.

Harold Innis too spoke to this insatiable quest for resources as he outlined Canada's wilderness economy in *The Fur Trade in Canada*. Although a few of his ideas are seen by some economists today as controversial, there are times when reading this 1927 book that it seems like it was written just last week. Replace "fur" with "oil" or "water" or "timber" and Innis's canonical work becomes current yet again. It was not just fur that fed the fur trade: every fort needed trees to build it and provide heat; every dog team needed fish or some form of meat to fuel it; every voyageur needed pemmican or other staples to keep his arms paddling; and every Red River cart needed feed for its oxen to keep its wheels turning. The land always provided, and until more permanent extractive activities were developed, it was able to recover either fully or at least with the right shade of green to camouflage our questionable ecological history. Perhaps it is this ability of the land to rebound that has lulled us into a state of complacency of sorts. It may also be that, for the first time in the relatively young histories of both Canada and the United States, more than 80 per cent of our populations live in urban areas. In Canada, most of these cities are snuggled up close to the 49th parallel, far away from any impacts in what we call "the wilderness."

Through mechanization, electronic distractions and easy access to necessary resources, we are becoming increasingly disassociated from our environment. It is a state of "nature deficit disorder" as described by Richard Louv in his book *Last Child in the Woods*. It is also a phenomenon I observe every semester in my own classrooms. In my introductory course in environmental science, I begin the first class with a quick quiz. Students are asked to identify a series of twenty images: ten portraying things commonly found in the natural environment (an aspen tree, a black-capped chickadee) and ten from popular culture (an electronic device, a television character). I make sure that most of the natural objects are ones students would regularly encounter in cities. Some of the images from popular culture include relics such as a typewriter (circa my own era) or politicians and celebrities (Barack Obama, Marge Simpson and the Mini Cooper always beat out Stephen Harper). Regardless of the class, images from popular culture are identified correctly at least 90 per cent of the time, while those from the natural world languish at only about 30 per cent correct. For electronic items such as the iPod or the Nintendo Wii, I am often even provided with technical specifications such as which generation of device it is, its memory capacity and so on. For the natural images, it is ones that appear in television documentaries (e.g., elephant seals) that are sometimes more recognizable than, say, the wood frog that sings people to sleep every night in the spring. The frog is recognized less than 2 per cent of the time even though it is one of the most common and broadly distributed amphibians in North America. These classes consistently have between sixty and eighty students, so sample size is not an issue.

One of Louv's main arguments is that we are unlikely to protect what we do not know. How we manage our wildlife will be reflected in how well we understand it and respect it. My job as a park warden was to help protect what I knew and was raised to love and respect. My goal as an educator and researcher is to teach others to do the same, through my stories and knowledge and their experiences in and outside the classroom. It is society as a whole that needs to see how animals such as beavers modify landscapes and bring life back into arid environments. People need a renewed appreciation of how the waters these amazing creatures provide allow for all those other stuffed, chocolate-dipped and appliquéd symbols of Canada to survive and flourish as well. Ironically it is the "strange man sitting by the side of the river" – the naturalists, biologists and hunters – who are the ones who see beavers gliding through the water on a perfect summer's evening and who know the land the best and can fight for its protection. Yet it is the average urbanite who glibly speaks of his or her innate connection to the wilderness and love of all things furry, despite never venturing beyond the edge of the city. As they cheered the giant inflated beaver and moose at the Vancouver Olympics, I was sadly aware that so few of those people had ever seen either of these common animals in the wild.

We are still consumers of natural resources, and by our nature as human beings we always will be, in order to survive. But how we manage those resources is critical. Loving nature to death and coveting its symbols while denying its needs is not a viable approach. To manage the conflicts between wildlife and humans requires an experiential understanding of our wild neighbours and the creativity, time and adaptability to do the right thing while we still have a chance. Such creativity is slowly emerging in the management of beavers and people, and these innovative approaches provide hope for a more harmonious future for all of us.

Battle of the Droughts

When a wetland goes dry, some only see a loss of water. For its plants and animals, water loss is a loss of critical habitat and often can mean the difference between life and death. Many plants and even some aquatic invertebrates can maintain dormancy in the soil for several decades in dry conditions, but eventually the battle is lost if water never returns. We are now facing a scenario where we could easily have to deal with more frequent and severe droughts in many places in the world. Beavers have endured extremes in climate for millennia, but for beavers and humans alike in this current era, it could be a time like no other.

In their paper "An Impending Water Crisis in Canada's Western Prairie Provinces," ecologists David Schindler and William Donahue indicate that a combination of climate change and the modification of the natural environment by humans has already brought us into a new reality. The rivers flowing from the icefields and glaciers high in the Rocky Mountains already have reduced flows, which translates into a "pending water crisis" for the Prairies, which depend on these ancient sources of water. The summer flows in the South Saskatchewan River have already decreased by 84 per cent and the glaciers themselves have shrunk by almost 25 per cent. In many parts of Canada and the United States, temperatures have increased by 2°C since the 1970s and continue to increase. The predictions are for hot and dry with a bit of arid thrown in for good measure.

As the glaciers diminish, the rivers they feed will become a fraction of their former size. This is not welcome news for cities such as Calgary, Edmonton, Red Deer and Saskatoon, or any of the species that call these rivers home. Like the beaver, these seemingly ever-present glaciers, which framed so many honeymoon and holiday photos of the past century, have been around since the last great Ice Age. The climate has been up and down for the past ten thousand years, but the accelerated rate of global warming may mean that even a long cool-down must come to an end.

When I first started with Parks Canada in 1988, a hike up Wilcox Pass near Banff and Jasper's Columbia Icefields was a must. So I hiked up the steep but lovely trail to a beautiful view of Athabasca Glacier, Snow Dome and the mere sliver of the vast Columbia Icefield (largely hidden by adjacent mountains) that was mother to them all. Click went the camera and off I went to explore some more alpine terrain. Almost twenty years later, just after finishing my career with Parks Canada, I hiked up the trail again and could not resist the classic photo opportunity. When I got home, I happened to compare the two images and was shocked at the amount of ice that had melted since my original photo was taken. The glaciers were not just shrinking in length, they were actually thinner than before, too. We will need more than a colony of beavers to stop what is to come during this period of a rapidly warming climate, but given what I know from my own research and that of others, it sure would be a benefit to keep them around.

Drought is nothing new to the Canadian Prairies. The Dirty Thirties changed the practices and psyche of the nation forever. Widespread crop failures, dust storms arising from poor land management, and swarms of locusts devastated the Canadian economy during that time. More than a third of the population was employed in agriculture for much of the first half of the 20th century. More recently, statistician Geoff Bowlby's article "Farmers Leaving the Field" shows a dramatic decline in the number of farms and full-time farmers in Canada since the Second World War, despite a dramatic increase in the size of farms. Even without drought, it has often been difficult to make a go of it with farming. In reality the loss of the family farm is as much an effect of changing social patterns relating to labour and globalization as it is a matter of climate. For those farms that remain, however, the more recent drought that reached its peak in 2002 has had devastating results across the Canadian Prairies. Some of my friends who owned farms and ranches never recovered and almost ten years later still suffer the emotional and financial costs of that drought.

What is ironic about the last major drought on the Canadian Prairies, from 1998 to 2004, is that many of the older farmers who had lived through the 1930s managed to keep at least one active beaver colony on their property as a safety net. Regardless of having forgone the arable land they could have gained by draining such wetlands, these quiet champions of the beaver knew full well that during dry times, keeping those ponds would result in a net benefit to the health of the farm. Beavers know how to keep water around. Their ponds mean water for livestock, pasture for cattle and easier times during the lean years. It is a lot easier than excavating dugouts, damming streams and trying to replicate what a beaver does for free.

A *New York Times* article from 1902, "The Beaver's Extinction," tells the story so well.

The anonymous author depicts a conversation with an old scout who describes the beaver as a "weather specialist." Not only could beavers predict a hard winter in the middle of summer, said the scout, they could also predict drought. It was through the "hot winds drying and burning everything up," he explained, that the beavers could foretell the dry weather to come and would dig a well "twelve to fifteen feet deep" under the roots of a tree ("to avoid buffalo"), thereby securing water throughout the drought. Apparently, the thirst-stricken buffalo would seek out these beaver wells and, in their quest for water, fall into the holes and become "the only trouble for the beavers." It is a touching piece of journalism that was a final plea to protect some of the few remaining beavers in Routt County, Colorado. Whether the author succeeded is unknown, but his passion certainly reveals a true champion for the cause.

One can suppose that beavers got a lot of practice reading those "hot winds" during their long evolution. Approximately 1,200 years ago, the earth entered a drier period that peaked during what is known as the Medieval Warm Period and ended during the 13th century CE. The general belief is that the warmer temperatures allowed ice-free travel across much of the North Atlantic and likely allowed the Vikings to land in North America during this period. The beavers, as always, were there to see it happen. It also represented a much drier period in the earth's more recent history. In reality, as Schindler and Donahue point out, the 20th century was one of the wettest periods during the past two millennia. They also determined that dry conditions are more likely the norm for Canada's western prairie provinces and that several droughts lasting many decades were quite common in previous centuries. The aridity of the Dirty Thirties, in their opinion, was actually mild relative to historical norms.

Although global warming still brings on keen debate in some ever-diminishing circles, there are very few people who disagree that things are getting a little odd on the climate front. As I sat through another Alberta winter with January temperatures well into the negative double digits, I heard a weather report of rain in Iqaluit, the capital of Nunavut and one of Canada's most northerly cities. Iqaluit is generally in the low minus-thirties at that time of year. How the global landscape will change as the climate warms is the focus of many ecological and socio-economic modellers and mathematicians. How wildlife will react is less certain, although some trends are already emerging. Species such as white-tailed deer, coyotes and raccoons are already moving north to take advantage of land clearing and warmer climates. American robins are now nesting in Canada's High Arctic, something that was unheard of less than twenty years ago.

Along with other studies, Suzanne Bayley's and my research showed a distinct benefit in beavers' ability to mitigate the effects of droughts. But data sets that extend over thousands of years, as reflected by the research of Lyman Persico and Grant Meyer, suggest that even beavers have their climatological limits. In a 2009 article, "Holocene Beaver Damming, Fluvial Geomorphology and Climate in Yellowstone National Park, Wyoming," the authors make it clear that despite the ability of *Castor canadensis* to have identifiable impacts on landscapes for thousands of years, the beavers curbed their desire to build dams, and their presence on the landscape was less apparent, during extensive dry periods and times of unstable climate. Unlike our study that extended over five decades, Persico and Meyer's data spanned four millennia. It seems that the beaver is not completely infallible and might need some help when it comes to fighting the effects of global warming and declining water resources. Yet regardless how humans fare, there is no doubt that, barring a resurgence of a fur trade, beavers will engineer a way to remain on much of the North American landscape. Whether land managers can bring themselves to see beavers as allies rather than pests, however, is as unpredictable as the weather.

Furry Vision

So why all the conflict between beavers and modern humans? My theory is that two control freaks will battle over the same tree until long after the last of its stump decays back into the forest floor. Humans simply do not like to be outdone by a rodent, plain and simple. The historical record, however, shows that not only the beaver but also the Norway rat and the house mouse have almost always won the war. Although cockroaches are touted as the most persistent animals on the planet, they were never chased down for their furs, turned into hats or marketed as a perfume. The evidence is in, and rodents rule the world.

Subliminally, Canadians have always known this, as noted in the legal recognition of the beaver as Canada's national symbol with the

National Symbol of Canada Act in 1985. The purpose of this law is "to provide for the recognition of the Beaver (*Castor canadensis*) as a symbol of the sovereignty of Canada." The Act contains only two sections: the short-title statement, and section 2, which reads:

> It is hereby recognized and declared that the Beaver (*Castor canadensis*) is a symbol of the sovereignty of Canada and it is proclaimed that any representation of the Beaver (*Castor canadensis*) when used by Her Majesty in right of Canada shall be so used and so regarded.

Powerful words for a country built on the felt of these watery rodents. Read a certain way, they almost sound a bit like reconciliation. But a complete truce has yet to come. In reality, it was the relentless quest for beaver pelts that in large measure allowed Canada to develop into a sovereign country in 1867. Throughout Canadian history, the beaver has graced our money, the first postage stamp, and the coats of arms of both the Canadian Pacific Railway and the Hudson's Bay Company. It has served as the symbol for Canada's national parks, been the mascot of many sports teams and the 1976 Summer Olympics in Montreal, and is fittingly displayed as the symbol for the Faculty of Engineering at the University of Alberta. The list of beaver affiliations could also include airplanes and movie stars. Visual representations of the beaver are everywhere, from downtown Toronto to a 4.5-metre-high, 1300-kilogram sculpture of the world's largest beaver at, where else, Beaverlodge, Alberta. It is even the representative animal of the State of Oregon, a place where it faced some of its greatest challenges during the fight between HBC and the American fur companies.

One of my most recent beaver gifts is a toque made of New Zealand wool. It looks like a beaver's head, complete with tiny ears, a cute little nose and big brown eyes. When I wear it on my winter walks, I am keenly aware that any beaver strolling outside its lodge during winter would be the next meal of the nearest coyote, cougar or wolf. More than reinforcing the idea of our vulnerability to northern winters, though, this simple toque reminds me of how we interact with our natural world and how tenuous that relationship really is.

In his book *Biodiversity in Canada: Ecology, Ideas, and Action*, Stephen Bocking provides an excellent journey through our country's evolution of its ecological identity. Following the peak of the fur trade and the settling of Canada, Bocking describes how "railways, guns and fences" created the prairie landscape we see today and how saws and axes staged battles with trees in the forests of Ontario and points east. In other regions of colonial Canada, the relationship with the wilderness was one based on distrust and even fear. These European settlers found the vast untamed forests to be frightening and even thought that civilized men who went into the wilderness would become wild and lose touch with their former selves. For earlier settlers, wearing a beaver hat was to keep warm, not to covet the cute physical features and symbolism of the rodent itself.

It was not until the mid-1800s that a growing economy and prosperity of nationhood would ignite an interest in nature and environment. The country was in its infancy and its sheer geographical size reinforced the idea that wildlife was "inexhaustible" and flourishing on the "frontier." Following similar ideas in the United States by Henry David Thoreau, Ralph Waldo Emerson and John Muir, there was a "back to nature" movement that resulted in an invigorated crop of hikers, campers and bird-watchers. Nature appreciation turned a fear of wilderness into respect and created a core part of the Canadian identity. Following the establishment of the world's first national park, Yellowstone in 1872, the first Canadian national park, Rocky Mountains Park (later known as Banff National Park), was established in 1885. Several more parks followed in the next few years and their popularity grew accordingly.

Unlike the United States, however, the concept of wildlife and resource conservation was slow to come to the "land of the silver birch, home of the beaver." As Janet Foster points out in her compelling book *Working for Wildlife: The Beginnings of Preservation in Canada*, wildlife was still considered unimportant by governments and citizens alike at the turn of the 20th century. As seen in many early government reports, the importance of wildlife was measured by the revenues it produced. Even in our national parks, "bad animals" such as wolves and cougars were killed, while "good animals" such as deer and elk were left to wander undisturbed. Naturalist clubs were local entities only and not prone to lobbying governments. In Canada, governments were focused on natural resources and parks as a means of revenue generation rather than nature preservation or conservation. Even Banff National Park was created, in part, to help pay for the completion of the railway, while Yellowstone was set aside to protect its outstanding scenery and natural values from the extensive resource exploitation seen elsewhere in that country. Economic development through resource exploitation was the primary focus of the government of Sir John A. Macdonald, and in many ways it still underlies Canada's economy.

Foster's book attributes the awakening of the public and their demand for government action in Canada to the devastation and finally the complete extinction of the passenger pigeon due to overhunting. By this time, beaver, elk and bison had already declined to near extinction in many parts of the country. An 1892 Royal Commission on the state of wildlife populations in Ontario determined that habitat loss, "ravages of wolves," overhunting and overfishing were "indeed a deplorable state of affairs." Laws were passed and policies were drafted.

Parks, such as Algonquin Provincial Park (established 1893), were developed and legally protected, albeit with shared use. Commercial logging in the park continues today. Early parks in Canada were designed for multiple uses and, in the case of national parks, revenue generation. Today they serve as much an ecological function as an economic one, though the tendency to tie them into a business model of administration and "cost recovery" always exists as part of the bottom line. Despite the economic side of our protected areas, laws such as the Canada National Parks Act offer full protection to natural resources within park boundaries.

Legislative solutions may be helpful if they are based on good wildlife science, which can be scarce enough even today and was in its infancy for much of the 20th century. Even if the laws are good ones, they must be enforced. Governments can write all the legislation they want, but if there is no commitment to providing people at the field level to enforce them, they are just words without intent. Changing a national psyche from one of complete access to resources to one of limited or no harvest would take time. Garrett Hardin's famous paper "The Tragedy of the Commons" (1968) highlights how the independent actions of multiple individuals negatively affect not only other people's access to resources but also the resources themselves. In the end, depletion is inevitable and short-term gains for the individual replace long-term benefits for all. Two of the solutions he suggested for managing the commons are to regulate access or allow private ownership. Hardin's ultimate solution – stopping human population growth – still has not caught on, although attempts have been made in the darker side of history.

We have seen this scenario of resource exploitation repeat itself in Canada with the depletion of the cod fishery, the near-extinction of the beaver and plains bison, the complete annihilation of the passenger pigeon and the great auk, and the loss of much of our native prairie habitats and eastern forests. Our past resembles a horde of Pac-Man-like colonists who moved across the landscape eating everything in their path. We are marginally more conscientious today, but global warming and overexploitation of water resources pose new challenges for the commons. Whether we rise to these challenges, and turn tragedy into triumph, remains to be seen.

While Americans followed the passions of conservationists such as John Muir, Aldo Leopold and even a president, Theodore Roosevelt, Canada had its own suite of far-sighted thinkers. Today, Canadians benefit from the visionary actions of several government officials during the beginning of the 20th century. They include people like the first national park superintendent, Howard Douglas; the commissioner of the Dominion Parks Service from 1911 to 1936, James (Bunny) Harkin; and Gordon Hewitt, a Dominion entomologist and consulting zoologist to the federal government who helped spearhead game legislation, including a major international treaty and its associated law that still protects migratory birds today. Granted, more Canadians will know the names of the three Americans mentioned and only a meagre few will recognize the likes of Douglas, Harkin and Hewitt, but collectively they were champions of wildlife conservation in Canada. In the case of Harkin, his work facilitated the accomplishments of many others, including Grey Owl and Anahareo.

Their journeys were not without mistakes and controversy, but their positive contributions were long-lasting and beneficial to all.

Despite our country's exploitative history with natural resources, there is still resounding support for our protected areas and wildlife in North America. Many of our national parks welcome tourists throughout the year now. In the 1990s, there was talk of enticing more visitors to the parks during the "shoulder seasons" – those quiet times between summer and ski season (fall) and between ski season and summer (spring). As usual it was an idea based in revenue generation, but there were whispers among Parks staff that perhaps even more Canadians would be turned on to natural areas and their protection. The concept worked and visitation to many of our parks has in fact increased (although overnight backcountry use is on the decline).

While I was working in the Rockies, three animals were definite bucket-list material for park visitors: bears, moose and beavers. Although tourists had all sorts of fun putting their children on the backs of bighorn sheep (believe me, it would happen – I was the first responder) and petting elk calves (not a good idea without asking mamma elk first), seeing one of the three iconic Canadian animals meant they had finally found the wilderness. Bears and moose were sometimes a more difficult find, but beavers were a sure thing if there was an active lodge nearby. Sending the visitors out at dusk to the edge of a pond inevitably resulted in ear-to-ear grins that were unstoppable.

The beaver represents more than wilderness. It is a symbol of hard work, tenacity, duty – all values strongly supported by Canadians. Yet it is nothing short of a miracle that there are any beavers left to see.

Emerging approaches to beaver manage-ment may help keep tourists and locals alike in awe of that glistening patch of fur gliding through the water. In many areas, specialized flow-control devices have been installed to allow water to be drained from beaver ponds quietly so that the beavers don't hear the sound of running water that drives them to action. They remain none the wiser and are not inspired to plug the culvert and raise the level of their pond to new heights. The ponds stay at stable and ecologically valuable levels, happy beavers remain on the landscape and infrastructure such as roads and trails remain intact. A win-win scenario all around.

What does this "new" approach to management mean in dollars and cents? It has never really been studied to any great extent, although some of these "new" approaches have been around since the 1960s. What is known, however, is that battling beaver activities the traditional way costs hundreds of thousands, if not millions, of dollars per year. A report by Elaine Menzies, "Cooperative Beaver Management in the Riding Mountain Biosphere Reserve, Manitoba," and another paper by Paul Jensen and others in the *Wildlife Society Bulletin* reported that compensation to trappers and facility repairs (e.g., culverts and roads) can amount to $125,000 per year (at $15 per beaver confirmed killed) and up to $4,900 per incident, respectively. Both studies are more than ten years old now, and one can assume that costs have only increased. In my experience, the costs of managing conflicts between humans and beavers are poorly documented if at all.

Control devices, such as the Castor Master® and Beaver Deceiver®, developed by Vermont biologist Skip Lisle in the 1990s, are increasingly in demand as a more cost-effective solution. Once installed, these require minimal maintenance and can last for years with little trouble. A similar design we installed in Elk Island National Park dramatically reduced conflicts and flooding of a key road. When I chatted with Skip in 2009 at the Fifth International Beaver Symposium (yes, there are entire conferences about beavers – why did no one laugh at me when I went to conferences about bears?), he mentioned having been flown to Poland earlier that year to install a Beaver Deceiver there. Just as fur traders made their living by removing beavers, Skip and others like him now make their living keeping them on the land.

A more holistic approach to wildlife management makes sense. It is most likely cheaper in the long run, more ecologically sound, and works with the biology of the animal rather than against it. And let's face it, a bunch of grown men chasing beavers around with dynamite, backhoes and five or six assistants just looks silly – especially if all this activity is done at the same locations year in and year out with exactly the same outcome. The maxim that "insanity means doing the same thing over and over again but expecting a different result" seems to apply very well to past forms of beaver management. There is no doubt that we need a new approach. Changing tack would not only aid the many species that depend on the wetlands beavers create, but it would help government coffers too. Perhaps we persist in battling beavers as we always have because we simply cannot abide being outdone by a rodent with buck-teeth. I would like to think that by changing course, we would win much more than we have lost.

Hats Off to Tenacity

We will never know how the loss of millions of beavers changed the landscape over the decades during and after the fur trade. What is the ecological legacy left to us by Samuel de Champlain, George Simpson and the various companies that rose and fell with the fortunes of the beaver pelt? What we do know is that beavers and water are as inseparable as roots and trees. The landscapes of North America and Europe evolved with the beaver, and without it their ecology goes awry. Beavers represent the heart that pumps blood through the veins of our landscape. Without the ponds and channels they create and maintain, the lifeblood of the land dries up.

Yet beavers have survived. They fit the definition of tenacity, which means "to keep a firm hold of something, to be persistent and hold one's position." We are fortunate that we took on a capable adversary so many years ago. One with the tenacity and experience to survive the many challenges it has faced over the millennia. The beaver's hold on the landscape has suffered some slippage from loss of habitat, European settlement, industrialization, overharvesting and human conflict. But despite it all, beavers now exist in the tens of millions and fill almost all suitable habitats in North America. Through careless translocations, the North American beaver has even found success on other continents, much to the detriment of the local ecology.

In Europe, the re-establishment of *Castor fiber* throughout much of its former range represents the remarkable ability of beavers to recover from disaster, particularly with human assistance. There has been an awakening of sorts, at least in some people, and the ecological benefits of reintroductions are apparent. Water is one of those dynamic resources that everything needs. But when it comes in excess, such as across our roads or up to our doorsteps, the love affair ends and the battle with water and its ally the beaver begins anew. Perhaps adapting our management approaches will help change these patterns. One thing for certain is that the time will come when we cannot afford to drain yet another wetland or remove yet another beaver because of sheer frustration or habit. Our changing landscape and warming climate will demand more respectful treatment.

What we have experienced with beavers is not much different than our treatment of many of our "natural resources." The very term is often defined as "something that can be used by humans" rather than the more ecological view of an intrinsic worth regardless of human needs. Despite centuries of overexploitation of resources on a global scale, we still stumble our way into ecosystems and remove the working parts. It is like a mechanic wearing reading glasses for a job that requires bifocals. The land ethic based on "dominion . . . over all the earth, and over every creeping thing that creepeth upon the earth" conferred by the Bible, drove the exploitative development of the Dominion of Canada. It is a story that has unfolded on every continent since humans advanced their thumb and started to travel. Surely we can learn from our experience, and from mounting evidence of a breaking point, and begin to see our world and our position in it more broadly.

Part of Parks Canada's mandate is to manage and use national parks in such a way as to leave them "unimpaired for future generations." Many other land protection agencies have similar statements in their guiding policies and legislation. It is a theme that should be adopted at a much broader scale, one that considers entire ecosystems and our place within them, not above them. As climate shifts to a warmer and drier condition in many parts of the globe, we need our allies. It is all well and good to cherish our protected areas and reduce our immediate ecological footprint, but we have now found a challenge that extends beyond everyone's boundaries, including the limits of technological fixes and easy solutions.

As a species that has lived through ice ages, multi-decadal droughts, and the Pleistocene extinctions that took half of its brethren, the beaver has something to teach us about survival and becoming a productive member of our ecological community. We have an opportunity to reinvent our understanding of humans, wildlife and our shared environment. We are not only answering to ourselves; we are providing a more hopeful direction for others yet to come. For the most part, humans are accustomed to living in a world of abundant resources and predictable climates. The only thing that has really kept us from living within the constraints of our environment is ourselves and our defence of self-imposed political boundaries. With new climatic circumstances and scarcer resources, we may find ourselves in a very different situation.

Our consumption of resources appears to be insatiable at this point. New "must have" products and consumables are released onto the market at a frenetic pace, and much like the beaver hat they become an essential commodity for an increasing number of the world's population. When we see a product in a shiny plastic case, we rarely connect it with the land that was cleared to mine and extract the minerals and petroleum products behind that innovation. We are increasingly removed from the very environment that offers us everything we need to survive. Richard Louv called it "nature deficit disorder." With regard to water, I call it "beaver deficit disorder." Somewhere along the evolutionary path, we have forgotten that water comes from streams, food from plants and animals, and that we are fallible despite Botox and anti-wrinkle cream. We need to think like ecosystem engineers and carefully create our path to more sustainable resources and better choices.

Very few other animals have challenged our actions and behaviours as much as the beaver has. We plant a tree; beavers can cut it down. We build a road; beavers can dig right through the roadbed and turn it into a creek. We drain a landscape; beavers build a dam and bring the water back. There is something in that persistent drive to sustain water on the landscape that is a clue for our own survival as a species. Whether we take the time to learn from other species depends on our own adaptability and willingness to see our world and the resources within us in a new light.

Bookshelf

Allred, Morrell. *Beaver Behavior: Architect of Fame & Bane!* Happy Camp, Calif.: Naturegraph, 1986.

Bocking, Stephen, ed. *Biodiversity in Canada: Ecology, Ideas, and Action.* Peterborough, Ont.: Broadview Press, 2000.

Boswell, Randy. "Rare Map Shows Canada Overrun by Beavers." *Edmonton Journal.* October 17, 2010.

Bowlby, Geoff. "Farmers Leaving the Field." *Perspectives on Labour and Income* 3 (February 2002): 13–18.

Busher, Peter E., and Ryszard Dzięciołowski, eds. *Beaver Protection, Management, and Utilization in Europe and North America.* New York: Kluwer Academic/Plenum Publishers, 1999.

Collier, Eric. *Three Against the Wilderness.* Victoria, BC: TouchWood Editions, 2007. First published 1959 by Clarke, Irwin.

Dahl, Thomas E. "Wetlands losses in the United States 1780s to 1980s." Washington, DC: US Department of the Interior, Fish & Wildlife Service, 1990. Available online, http://is.gd/orG3hY, accessed March 25, 2011.

Dahl, Thomas E., and Gregory J. Allord. "Technical Aspects of Wetlands: History of Wetlands in the Conterminous United States." National Water Summary on Wetland Resources, United States Geological Survey Water Supply Paper 2425 (1997). Available online, http://is.gd/QGb0D8, accessed March 15, 2011.

Diamond, Jared. *The Third Chimpanzee: The Evolution and Future of the Human Animal*. New York: HarperCollins, 1992.

Denys, Nicolas. *Description geographique et historique des costes de l'Amerique septentrionale avec l'histoire naturelle du païs*. Tome 1. Paris: Claude Barbin, 1672. Available online, openlibrary.org [search title with only the one diacritic just as shown here], accessed March 15, 2011.

————. *Histoire naturelle des peuples, des animaux, des arbres & plantes de l'Amerique septentrionale & de ses divers climats*. Tome 2. Paris: Claude Barbin, 1672. Available online, openlibrary.org [search title without diacritics just as shown here], accessed March 15, 2011.

Federal, provincial and territorial governments of Canada. "Canadian Biodiversity: Ecosystem Status and Trends 2010." Ottawa: Canadian Councils of Resource Ministers. Available online (pdf), www.biodivcanada. ca/ecosystems, accessed March 18, 2011.

Foster, Janet. *Working for Wildlife: The Beginning of Preservation in Canada*. 2nd ed. Toronto: University of Toronto Press, 1998.

Hardin, Garrett. "The Tragedy of the Commons." *Science* 162, no. 3859 (1968): 1243–1248. Available online, www.sciencemag.org/content/162/3859/1243.full, accessed March 18, 2011.

Hood, Glynnis A., and Suzanne E. Bayley. "Beaver (*Castor canadensis*) mitigate the effects of climate on the area of open water in boreal wetlands in western Canada." *Biological Conservation* 141, no. 2 (February 2008): 556–567.

Innis, Harold A. *The Fur Trade in Canada*. Toronto: University of Toronto Press, 1999. First published 1927 by U of T Library.

Jenkins, Stephen H., Peter E. Busher. "Castor canadensis." *Mammalian Species*, no. 120 (June 1979): 1–8.

Jensen, Paul G., Paul D. Curtis, Mark E. Lehnert, David L. Hamelin. "Habitat and Structural Factors Influencing Beaver Interference with Highway Culverts." *Wildlife Society Bulletin* 29, no. 2 (July 2001): 654–664.

Kay, Jeanne. "Native Americans in the Fur Trade and Wildlife Depletion." *Environmental Review: ER* 9(2), Special Issue: American Indian Environmental History (Summer, 1985): 118–130.

Louv, Richard. *Last Child in the Woods: Saving Our Children from Nature-Deficit Disorder*. Chapel Hill, NC: Algonquin Books 2006.

McCormack, Patricia A. *Fort Chipewyan and the Shaping of Canadian History, 1788–1920s*: "We Like to Be Free in This Country." Vancouver: UBC Press, 2010.

Menzies, Elaine. "Cooperative Beaver Management in the Riding Mountain Biosphere Reserve, Manitoba." Master's degree practicum, University of Manitoba, Natural Resources Institute, 1998.

Morgan, Lewis H. *The American Beaver and His Works*. Philadelphia: J.B. Lippincott, 1868. Available online, www.archive.org, accessed March 18, 2011.

Morse, Eric W. *Fur Trade Canoe Routes of Canada: Then and Now*. 2nd ed. Toronto: University of Toronto Press, 1989. First published 1968 by Government of Canada.

Müller-Schwarze, Dietland, and Lixing Sun. *The Beaver: Natural History of a Wetlands Engineer*. Ithaca, NY: Comstock Publishing Assoc., 2003.

Persico, Lyman, and Grant Meyer. "Holocene Beaver Damming, Fluvial Geomorphology, and Climate in Yellowstone National Park, Wyoming." *Quaternary Research* 71, no. 3 (May 2009): 340–353.

Pollock, Michael M., Timothy J. Beechie, and Chris E. Jordan. "Geomorphic Changes Upstream of Beaver Dams in Bridge Creek, an Incised Stream Channel in the Interior Columbia River Basin, Eastern Oregon." *Earth Surface Processes and Landforms* 32, no. 8 (2007): 1174–1185.

Rybczynski, Natalia. "Woodcutting Behavior in Beavers (Castoridae, Rodentia): Estimating Ecological Performance in a Modern and a Fossil Taxon." *Paleobiology* 34, no. 3 (August 2008): 389–402.

Schindler, David W., and William F. Donahue. "An Impending Water Crisis in Canada's Western Prairie Provinces." *Proceedings of the National Academy of Sciences* 103, no. 19 (May 9, 2006): 7210–7216.

Seager, Richard, and Celine Herweijer. "Causes and Consequences of Nineteenth-Century Droughts in North America." Drought Research, Lamont-Doherty Earth Observatory of Columbia University, n.d., http://is.gd/5CWSfR, accessed March 15, 2011.

Seton, Ernest T. *Lives of Game Animals.* Vol. 4, Pt. II, Rodents etc. Garden City, NY: Doubleday, Doran, 1929.

"The Beaver's Extinction." *The New York Times*, August 10, 1902. Available online, http://query.nytimes.com/search [enter title; click "Search," then "All Results Since 1851"], accessed March 18, 2011.

Tournay, Audrey. *Beaver Tales: Audrey Tournay and the Aspen Valley Beavers.* Erin Mills, Ont.: Boston Mills Press, 2003.

Udall, Stewart L. *The Quiet Crisis.* New York: Holt, Rinehart & Winston, 1963.

About the Author

Glynnis Hood grew up in the heart of British Columbia's Kootenay Region. Since her first "real" job as a naturalist at the Creston Valley Wildlife Centre, she has worked in various protected areas throughout western Canada and into the Subarctic region and Boreal Plains. After four years as a park interpreter in Banff National Park, she spent the rest of her 19-year career with Parks Canada as national park warden. Through her years working in the backcountry of Jasper National Park, the diverse habitats of Waterton Lakes National Park, the expansive forests and wetlands of Wood Buffalo National Park and finally the aspen forests of Elk Island National Park, she has conducted monitoring and research on several species of wildlife. Glynnis completed her MSc work on human impacts on grizzly bear habitat availability, and her PhD research on beaver ecology and management. After a satisfying and rewarding career with Parks Canada, she is now an associate professor in Environmental Science at the University of Alberta's Augustana Campus. She currently lives in the Beaver Hills Region of east-central Alberta and has three beaver lodges as her closest neighbours.

Other Titles in this Series

The Insatiable Bark Beetle

Dr. Reese Halter

ISBN 978-1-926855-67-7

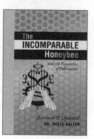

The Incomparable Honeybee

and the Economics of Pollination
Revised & Updated

Dr. Reese Halter

ISBN 978-1-926855-65-3

Becoming Water

Glaciers in a Warming World

Mike Demuth

ISBN 978-1-926855-72-1

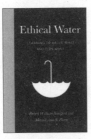

Ethical Water

Learning To Value What Matters Most

Robert William Sandford
& Merrell-Ann S. Phare

ISBN 978-1-926855-70-7

Little Black Lies

Corporate & Political Spin in the Alberta
Tar Sands

Jeff Gailus

ISBN 978-1-926855-69-1

The Grizzly Manifesto

In Defence of the Great Bear

by Jeff Gailus

ISBN 978-1-897522-83-7

Denying the Source

The Crisis of First Nations Water Rights

Merrell-Ann S. Phare

ISBN 978-1-897522-61-5

The Weekender Effect

Hyperdevelopment in Mountain Towns

Robert William Sandford

ISBN 978-1-897522-10-3

RMB saved the following resources by printing the pages of this book on chlorine-free paper made with 100% post-consumer waste:

Trees · 8, fully grown

Water · 3,886 gallons

Solid Waste · 236 pounds

Greenhouse Gases · 807 pounds

Calculations based on research by Environmental Defense and the Paper Task Force. Manufactured at Friesens Corporation.